Satellites in Space

Jill McDougall

Contents

Satellites in Space

What Is a Satellite?

A satellite is an object that **orbits** a large object in space, such as a planet. Some satellites are a natural part of our solar system. For example, the Moon is a satellite because it orbits Earth. However, most satellites in our solar system are artificial. These are satellites that have been built by people and **launched** into space from Earth.

An artificial satellite and the Moon both orbit Earth.

Today, thousands of artificial satellites orbit Earth, and they have a wide range of uses. Some can send TV, telephone, **GPS** (Global Positioning System) or internet signals to different locations around the world. Other satellites are used for scientific research, such as studying Earth's surface or looking deep into space.

Satellites can be built in surprising shapes and sizes. Tiny satellites, often shaped like cubes, can be small enough to fit inside a lunch box. On the other hand, the world's biggest satellite, the International Space Station, has giant wings and is roughly the size of a soccer field.

The International Space Station is the largest satellite in space.

The First Satellites

The first artificial satellite was launched in October 1957 by the **USSR**. This satellite, named Sputnik 1, was a simple metal sphere about the size of a beach ball. It had four long antennas to send **radio signals** back to Earth.

After three weeks in orbit, Sputnik 1's batteries ran out of power. The satellite continued to orbit Earth until it was pulled downwards by Earth's **gravity**. As Sputnik 1 sped through Earth's **atmosphere**, it burnt up.

The power for Sputnik 1 came from three batteries, which were kept inside the sphere.

In January 1958, the USA launched its first satellite, Explorer 1. This satellite was shaped like a small rocket and measured about 2 metres long. After its batteries failed, it remained in orbit for 12 more years.

A few months after Explorer 1 was launched, **space engineers** in the USA built a satellite that was fitted with solar panels to provide power. This satellite, named Vanguard 1, is still in orbit today, even though it no longer works.

The launch of these early satellites marked the beginning of a new age in Earth's history. This time is known as "the Space Age" – the time when humans first travelled into space and started exploring the universe.

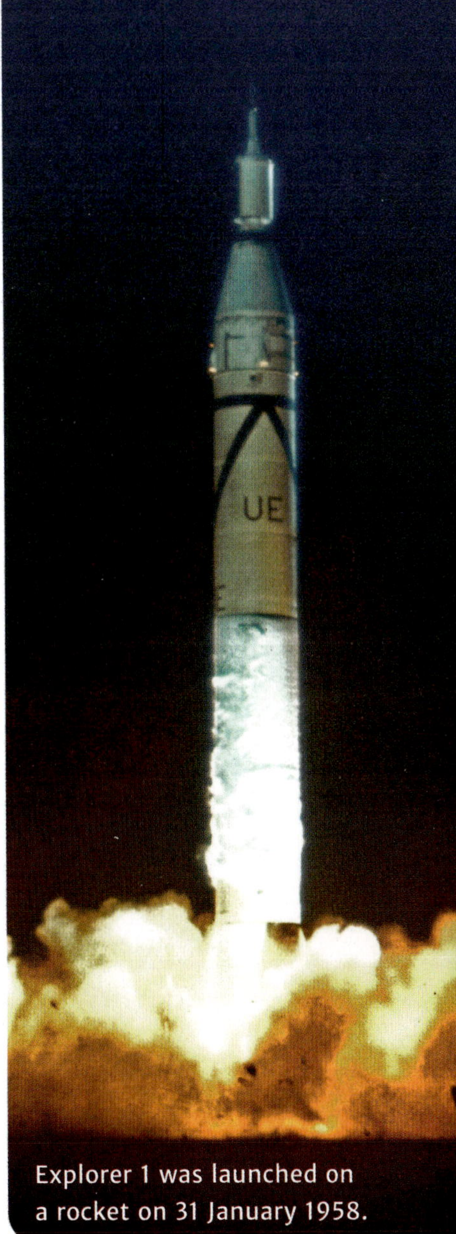

Explorer 1 was launched on a rocket on 31 January 1958.

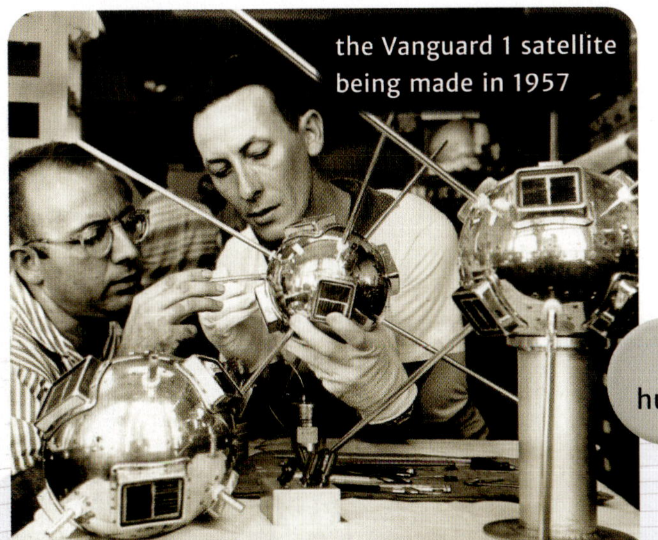
the Vanguard 1 satellite being made in 1957

Vanguard 1 is the oldest human-made object in space.

How Satellites Work

Satellites are usually launched into space using single-use rockets. Rockets need to use large amounts of fuel to propel themselves out of Earth's gravity and into space. Then, when a rocket reaches the correct distance from Earth, the satellite is released.

Once a satellite is in orbit, it can send and receive radio signals from Earth. When a signal is sent to a satellite, it can take a few seconds for the signal to reach the satellite and bounce back to Earth. This effect can be seen when TV interviewers talk to people on the other side of the world, as there is a short time delay between the question asked and the answer given.

This large satellite dish in Parkes, New South Wales, Australia, received TV signals of the first astronauts walking on the Moon.

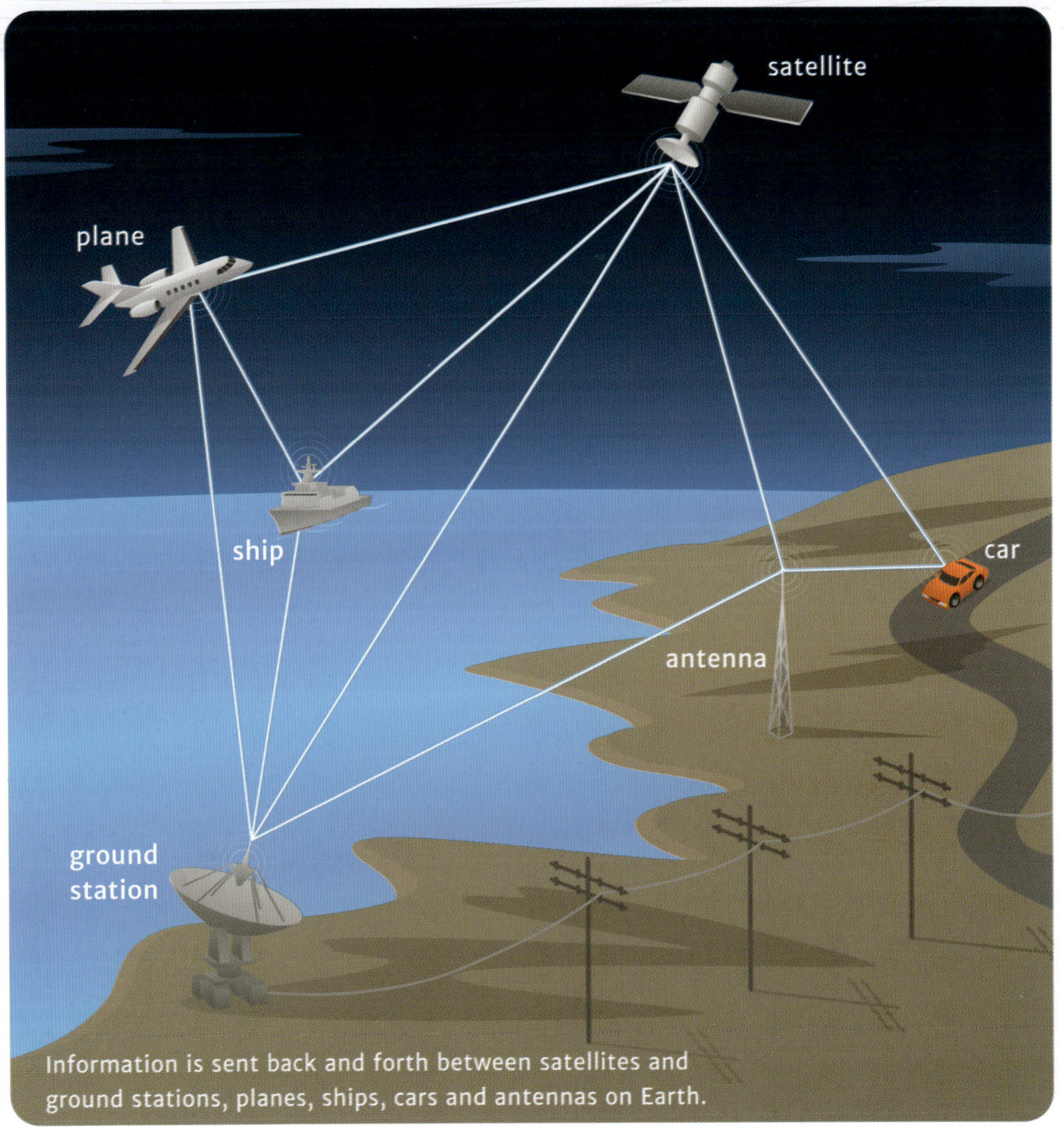

Information is sent back and forth between satellites and ground stations, planes, ships, cars and antennas on Earth.

Radio signals can be sent and received from a **ground station** using a large satellite dish. These ground stations are often built in areas where signals can be received easily, such as an open field or the top of a hill.

Satellites are launched to different distances from Earth depending on their purpose. Some orbit thousands of kilometres out in space, while others are just a few hundred kilometres above Earth's surface. The satellites that are closer to Earth can transmit signals at a higher speed because the signals have a shorter distance to travel.

The Ariane 5 rocket was used to launch the James Webb Space Telescope into orbit.

Satellites follow different orbits around Earth. Some travel over the North and South Poles, and others move around the equator. However, there are some satellites that always remain in the same place above Earth. They do this by travelling at the same speed Earth is turning.

A satellite is able to stay in orbit by balancing its speed with the pull of Earth's gravity. Without this balance, the satellite would either fly in a straight line, off into space, or fall back to Earth.

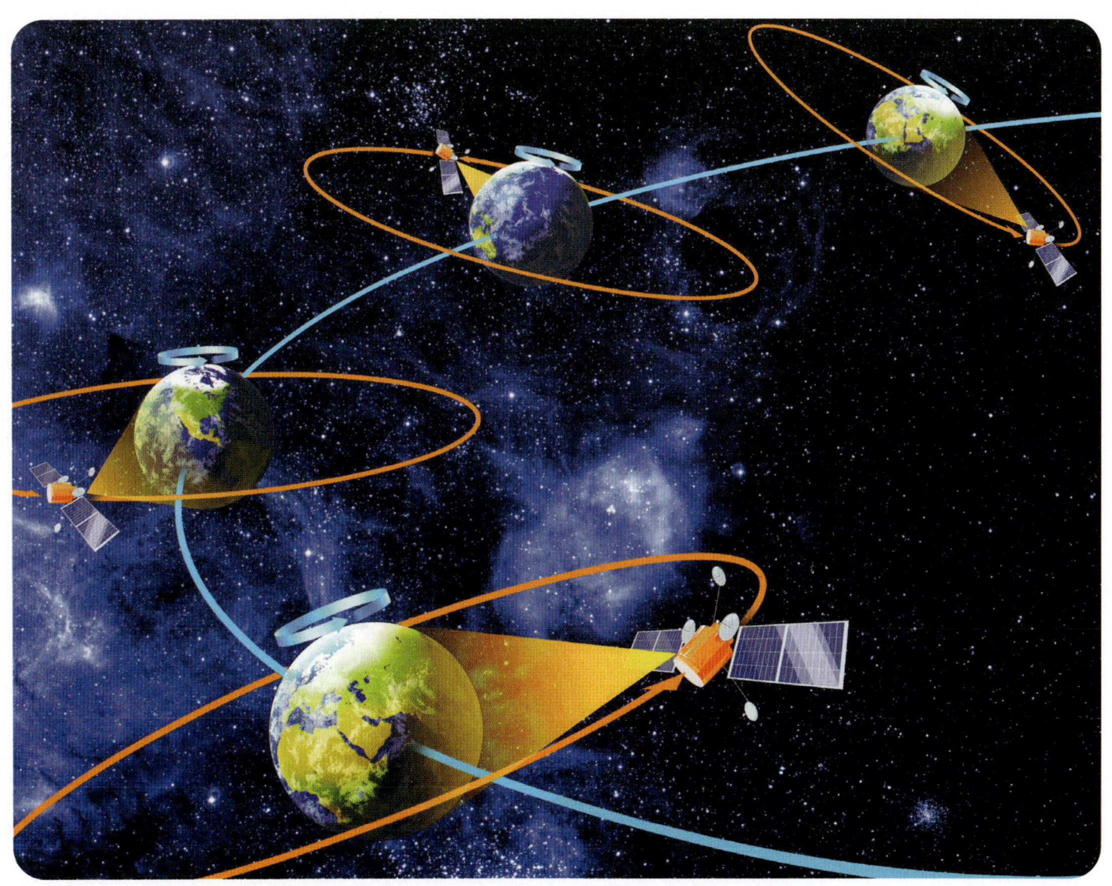

Some satellites are always able to view the same part of Earth because they orbit at the same speed Earth is turning.

What Satellites Do

In many ways, our everyday lives would not be possible without satellite technology. Some people depend on satellites to watch TV programs or access high-speed internet. Satellites also allow people to make mobile phone calls across the globe.

Satellites are often used to send signals to remote places that may be hard to reach in other ways. For example, a scientist in Antarctica or a tourist in the Australian outback can still search the internet or make phone calls by using signals that bounce off satellites.

A climber can make a phone call from the top of a mountain by using a satellite phone.

Some satellites help people **navigate** from place to place. When a person uses a map on a smartphone, they can see their exact location and receive specific directions. This is often called "using GPS" because the information comes from a group of GPS satellites.

GPS technology is also used to save lives. When people make emergency calls on their smartphones, their location can be quickly traced using GPS. In Aotearoa New Zealand in 2020, a teenager became lost while hiking in a national park. The search-and-rescue team was able to locate him quickly because he was carrying a GPS device that could give them his exact location.

GPS devices can help people know where they are when they're in the wilderness.

Some people fit GPS trackers to their pets so that they can locate them if they become lost.

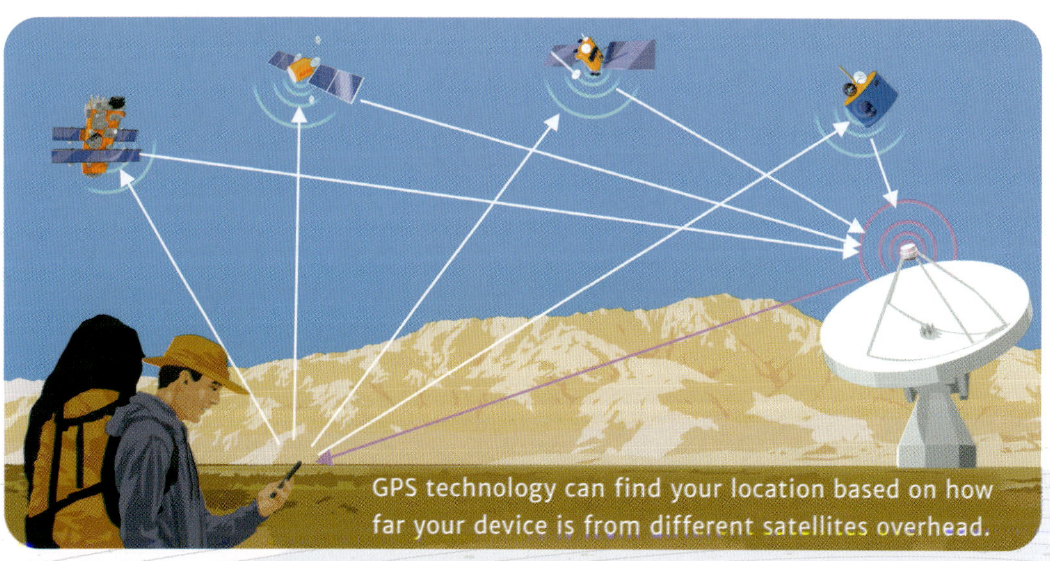

GPS technology can find your location based on how far your device is from different satellites overhead.

Some of the first satellites to orbit Earth were weather satellites. Today, weather satellites collect information about clouds, wind, storms, ice caps and much more. The information is then transmitted to computers on Earth to help **meteorologists** (pronounced *mee-tee-or-ol-oh-jists*) make daily weather forecasts.

Other satellites can provide early warnings for natural disasters, such as hurricanes and volcanic eruptions. To keep a watch on volcanoes, scientists use **sensors** on satellites to measure the heat coming from volcanic craters. If the temperature increases rapidly, scientists can send out a warning that the volcano is about to erupt.

A satellite took this image of a volcano in Tonga before it erupted in 2022.

One scientist, Dr Sarah Parcak, even uses satellites to find lost pyramids in Egypt. She studies satellite images to detect changes across Earth's surface, such as differences in rocks and plants, in order to find ancient structures buried underground.

Dr Sarah Parcak uses satellite images to study an ancient buried city in Egypt.

Satellites are also used to study space. These satellites carry huge telescopes that can collect information about moons, planets, comets and other galaxies.

Telescopes in Space

For many years, **astronomers** studied space by using powerful telescopes on Earth. However, Earth's atmosphere is made up of gases that make it difficult for scientists to get a clear view of the stars. Astronomers dreamed of sending a telescope into space that could capture clear images and help explain some of the mysteries of the universe.

The Hubble Space Telescope orbits Earth so it can take clear images of space.

In 1990, a satellite known as the Hubble Space Telescope was launched. This satellite, with its powerful cameras, enabled astronomers to make many important discoveries, such as how galaxies are formed. The telescope also captured images of moons orbiting Pluto that had never been seen before.

This 2004 image from the Hubble Space Telescope shows galaxies in deep space.

In December 2021, a new space telescope was launched. This satellite, known as the James Webb Space Telescope, is more powerful than the Hubble Space Telescope. It was also able to be sent further out into space. The cameras on the James Webb Space Telescope allow astronomers to capture detailed images of stars and planets that were formed over 13 billion years ago.

Scientists around the world can use the images taken by the James Webb Space Telescope to gain a greater understanding of the universe and how it began.

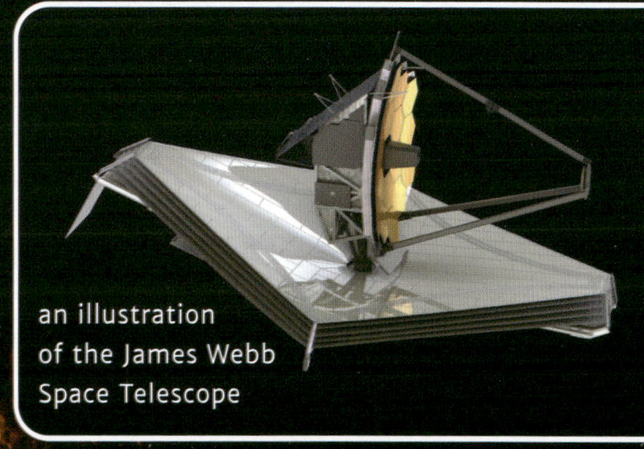

an illustration of the James Webb Space Telescope

The James Webb Space Telescope took this image in 2022 of the Southern Ring Nebula, a place where stars are born.

The International Space Station

The International Space Station is a satellite that was launched in 1998. It is the biggest single structure that humans have ever put into space. On a clear night, the space station can be seen as a bright, slowly moving light in the sky.

The space station is large enough for six or seven astronauts to live there for several months at a time. It contains science laboratories as well as bedrooms, bathrooms and a gym.

The International Space Station has large solar panels that help make electricity on board.

The space station's giant wings contain solar panels that convert sunlight into electricity, which powers the equipment on the station.

The astronauts' main role on the space station is to conduct scientific experiments. One of their tasks is to test the human body's ability to live in space. These tests help space engineers plan future missions to places such as the Moon and Mars.

The International Space Station flies 400 kilometres above Earth and completes one orbit every 90 minutes. This means that the station is in daylight for 45 minutes, then enters night for 45 minutes. The Sun rises and sets 16 times a day!

An astronaut from Japan tries to catch floating fruit delivered to the International Space Station.

Tiny Satellites

In the early 2000s, space engineers began to build miniature satellites that could be launched as a group. Many of these satellites are in the shape of 10-centimetre cubes.

When hundreds of tiny satellites work together in space, they can achieve important results. For example, together they can view every part of Earth's surface every day with their cameras. Scientists are then able to observe rapid changes in the environment, such as the direction of wildfires.

Sizes of Different Satellites

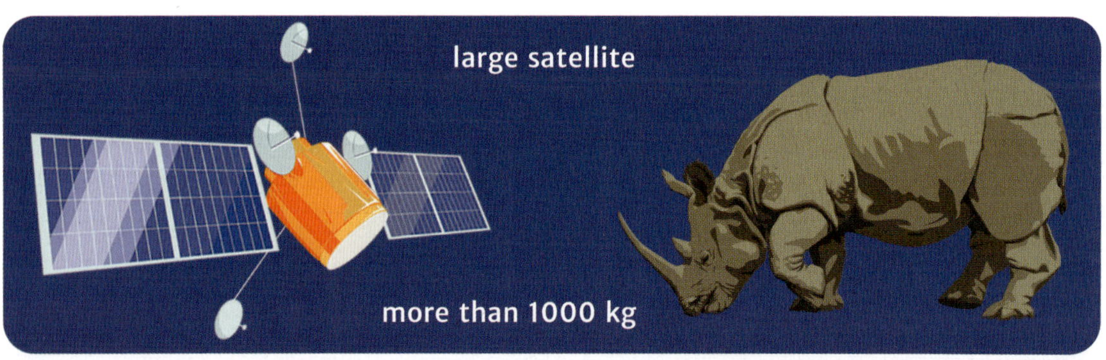

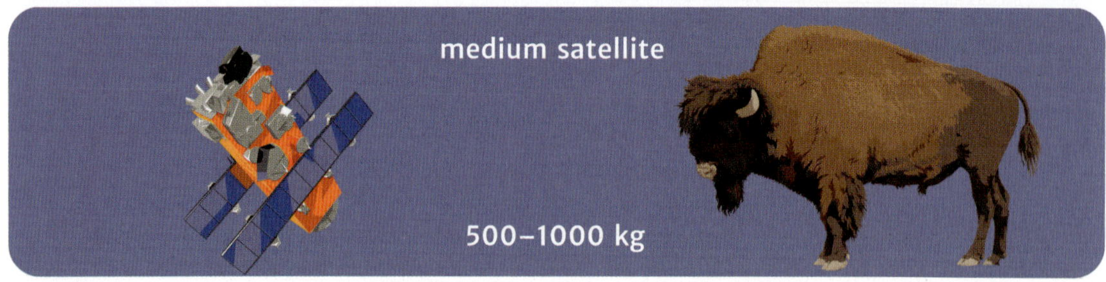

tiny satellite 1–10 kg

Although tiny satellites are very light, they can carry cameras and sensors, as well as small scientific instruments. In some schools, students have worked with space engineers to design and build their own satellites. These miniature satellites often carry out a small set of tasks. At a school in the USA, for example, 12-year-old students built a satellite that captured images of the effect wildfires had on the local landscape.

Students in California, USA, work to build a tiny satellite that will eventually take pictures of the Moon.

This tiny satellite was made in India.

Satellites and Earth's Environment

Satellites have become important tools in helping scientists and environmentalists care for the planet. The sensors that are attached to satellites can measure changes on Earth's surface, such as the temperature of the ocean, the thickness of the sea ice and the rate the glaciers are melting. This information allows scientists to study the impact of **climate change** around the world.

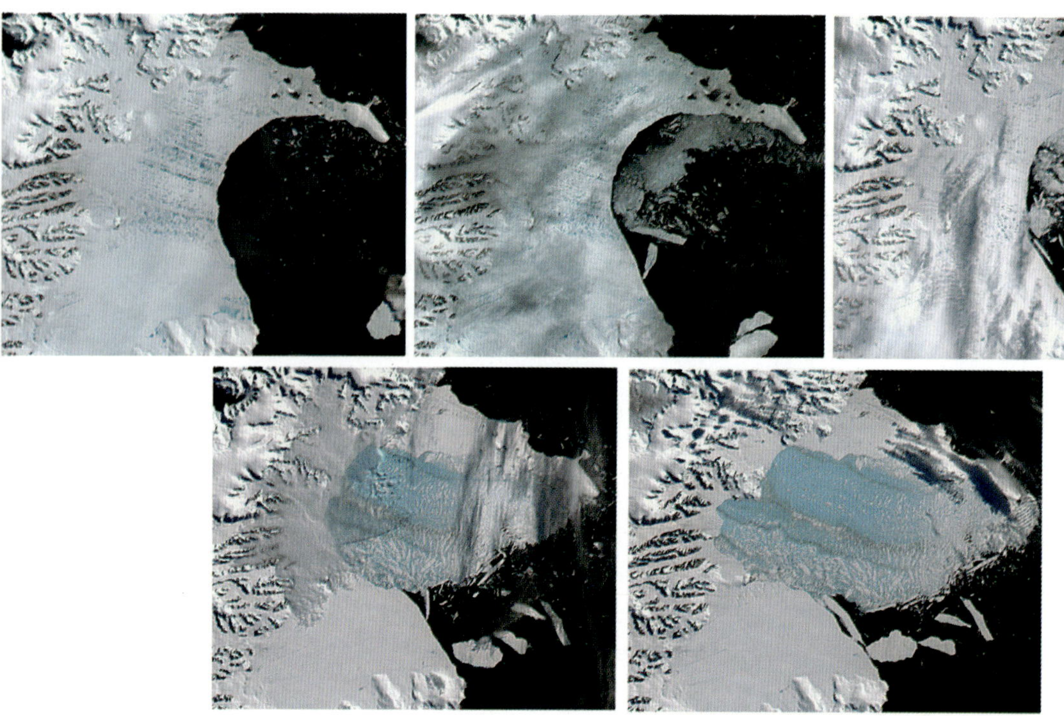

Satellite images show the collapse of an ice shelf over time: small pieces of ice break away, then large pieces, then partly melted snow slides down into the area.

Satellites can also be used to prevent certain activities, such as illegal **logging** in forests. Cameras on satellites can capture images of loggers building roads in forests before they begin to cut down trees. In this way, illegal logging can be stopped before it begins.

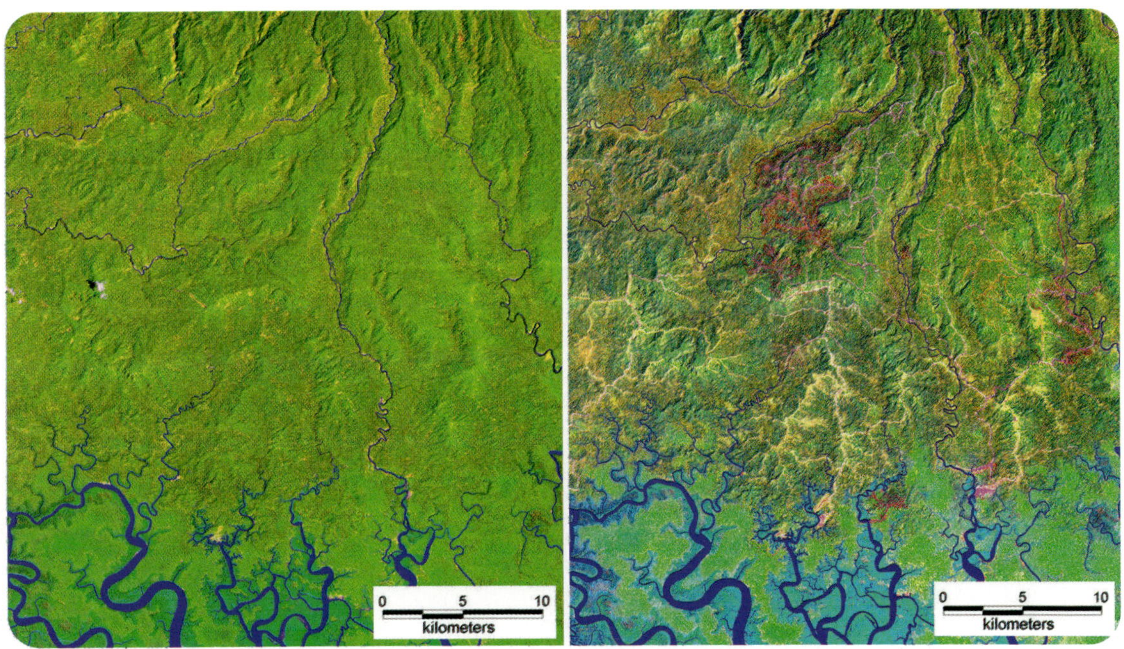

The amount of rainforest (shown in green) that has been lost in Papua New Guinea due to logging has been tracked using satellites, by taking photos of the same location over time.

In some countries, satellites are used to locate plastic rubbish on beaches. These satellites carry special sensors that can identify even tiny scraps of plastic among vast areas of sand and rocks. The information from the satellites is sent to local clean-up groups so that they can remove the rubbish.

Satellites at Risk

Since the beginning of the Space Age, thousands of satellites have been launched into space. In 2022, there were more than 8000 artificial satellites in orbit around Earth.

Since there are so many satellites, there is a risk of them colliding with pieces of **debris** called "space junk". The space around Earth is filled with this debris, which includes parts from rockets and old satellites that remain in orbit. Space junk can travel at more than 25 000 kilometres per hour, and when it strikes another object, it can cause serious damage.

In 2021, a tiny piece of space junk punched a hole in part of the International Space Station.

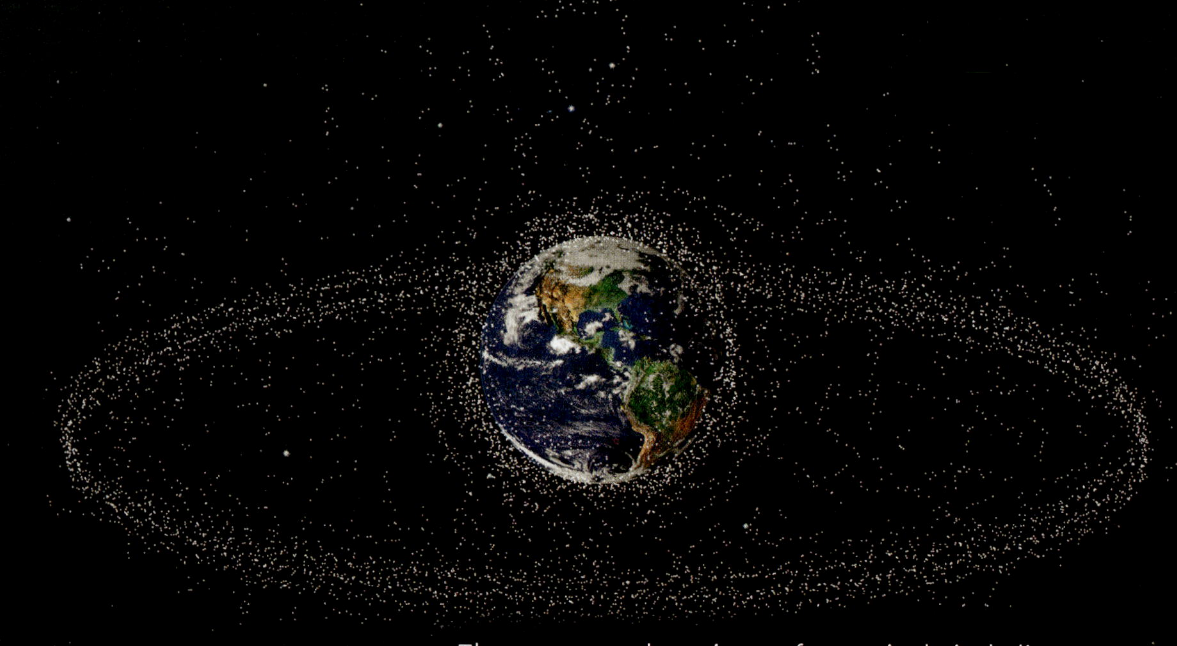

There are countless pieces of space junk, including satellites that no longer work, in orbit around Earth.

As space becomes more crowded, satellites are more likely to cross paths and collide with one another. In 2009, two satellites collided at a speed of more than 10 kilometres per second. They broke into thousands of pieces, and much of this debris remains in orbit.

Debris from colliding satellites continues to orbit Earth, creating even more space junk.

Satellites and the Night Sky

Satellites can look bright like stars because they reflect the light from the setting Sun. At night, satellites can look like a trail of white dots or streaks moving quickly across the sky.

As the night sky becomes brighter due to satellites, **constellations** of stars become less clear. Astronomers have found that the extra light from satellites makes it more difficult to capture clear images of the night sky.

An Australian First Nations astronomer, Karlie Noon, says that constellations are a vital part of First Nations peoples' cultural knowledge, and artificial lights make it difficult to observe them.

One important constellation is the Emu in the Sky, which is also known as the Dark Emu because its shape is made from dark patches in the sky, rather than from stars. Many Aboriginal cultures tell stories about the Emu in the Sky. For example, on Wiradjuri (pronounced *wuh-ra-juh-ree*) country, in central New South Wales, Australia, the emu's position in the sky signals the best time to collect emu eggs.

The Emu in the Sky constellation can be seen in the dark shapes between stars.

A trail of satellites can be seen lighting up the night sky in Canada.

It is difficult to imagine our everyday lives without satellites. However, many people are now becoming aware of the problems satellites can cause. Space engineers are designing satellites that will reflect less light in the night sky, and robots are being built to clean up space junk. As technology continues to improve, satellites are destined to play an even bigger role in our future.

Space Technology Can Change the World!

An Interview with a Space Engineer

Flavia Tata Nardini was born in Rome, Italy, where she studied to become a space engineer. In 2014, she moved to South Australia and started a space company.

Flavia Tata Nardini is a space engineer in South Australia.

Why did you decide to become a space engineer?

Ever since I was a little girl, I was determined to make a change in this world using space technology. I was fascinated by space and everything to do with it. I remember asking for space books and a telescope for my birthday. I thought space was super cool, and I wanted to learn more about stars, other planets and galaxies.

When it was time to decide what I wanted to study at university, I chose space engineering because I knew that space technology could help to make life on Earth better. Later, I started a space company that builds and launches **fleets** of tiny satellites.

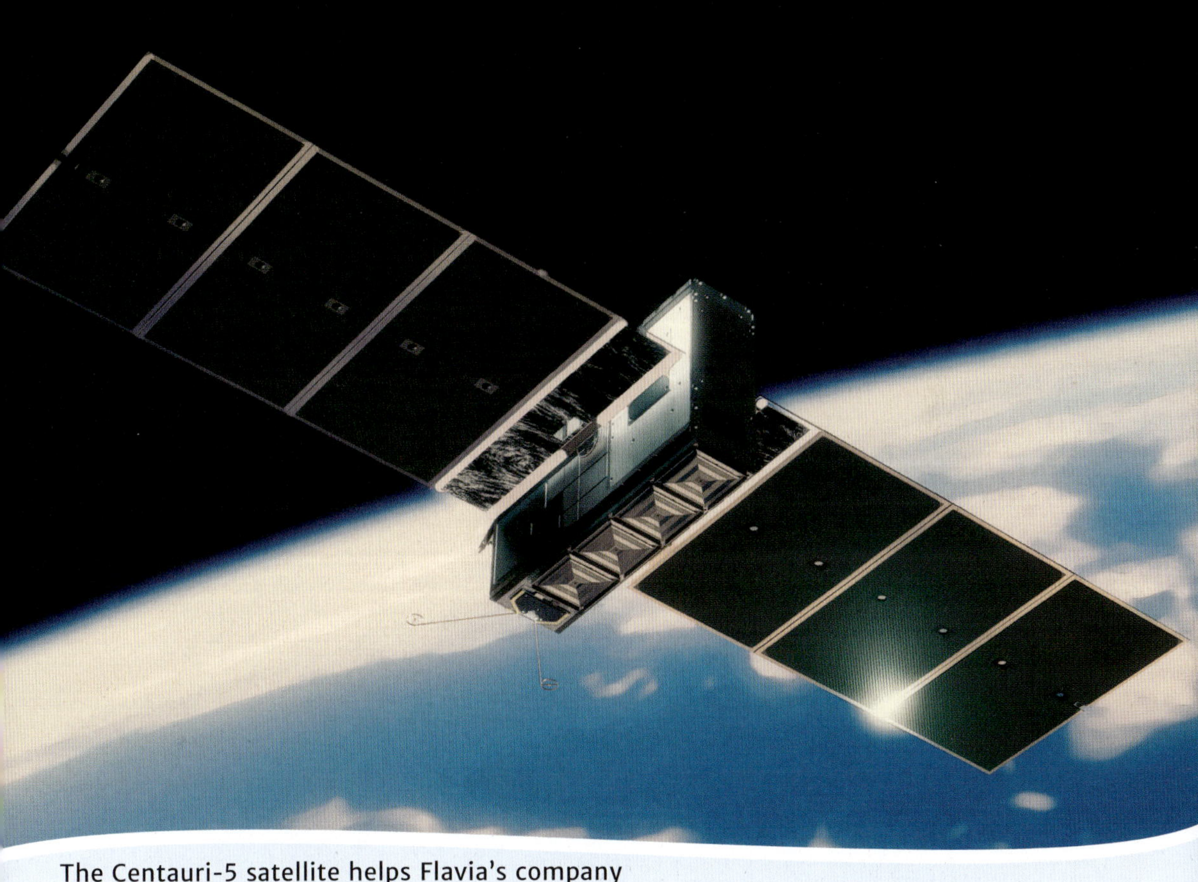

The Centauri-5 satellite helps Flavia's company search for minerals on Earth.

How can space technology improve life on Earth?

Space technology can solve many of our biggest challenges on Earth. For example, my company uses small satellites to search for minerals. These minerals are used in batteries for electric cars, and since electric cars reduce air pollution, that's good for us and the planet!

Using satellites also means that mining has less impact on the environment, because the sensors on the satellites can locate minerals as deep as two kilometres below the surface. The mining companies don't need to drill deep holes over and over again when they explore the land for minerals. We can tell them exactly where to drill.

Sensors that are planted in the ground help satellites to map a certain area.

a sensor used for mapping

Can satellites help us explore outer space?

Absolutely! Satellites help us explore Earth, and tiny satellites can help us explore beyond our planet. For example, in the future, satellites will be used to search for water on the Moon. If enough water is found, then the Moon could be used as a stop-off point on a journey to Mars.

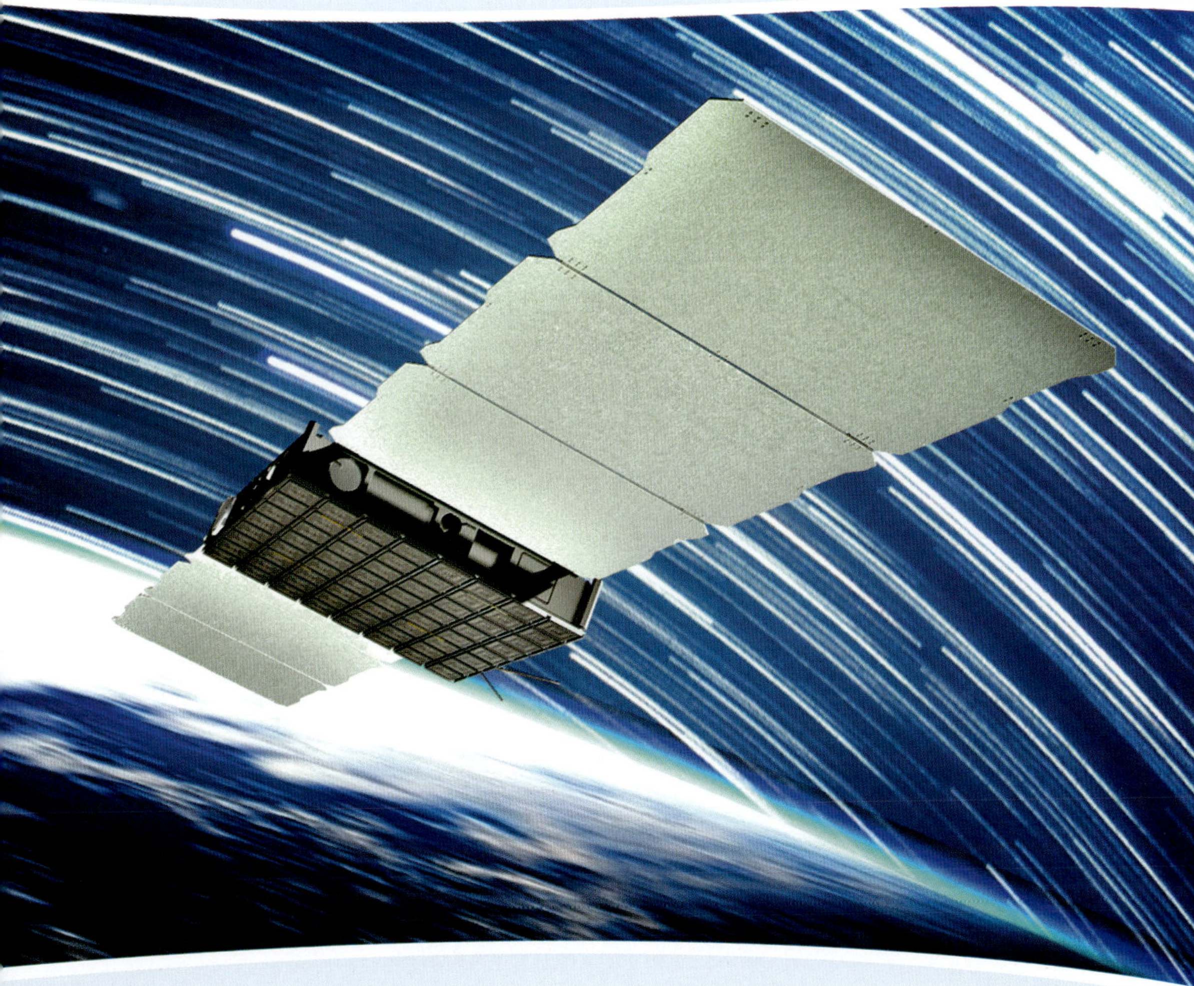

The Alpha satellite was the first satellite to be made with a 3D printer.

What advice would you give young people who are interested in a job like yours?

There is a big future for young people in space technology because there are many exciting jobs in this area.

My advice to a young person is this: follow your passion, and don't let anyone tell you that it's not possible. If you want to pursue a space career, you should spend your time discovering as much as you can about this amazing science.

I believe that everyone who works in space technology is an explorer, and if you are interested in space, you can be an explorer, too, and help change the world.

Anyone can study to become a space engineer and help build satellites in the future.

Glossary

astronomers (*noun*)	scientists who study space and the things in it, such as stars and planets
atmosphere (*noun*)	the layer of gases surrounding a planet
climate change (*noun*)	a global change in weather patterns
constellations (*noun*)	groups of stars that form patterns in the sky
debris (*noun*)	the remains of something broken
fleets (*noun*)	groups of objects controlled by one person
GPS (*noun*)	Global Positioning System: a navigation system that allows users to determine their exact location
gravity (*noun*)	the force that pulls all things on Earth towards its centre
ground station (*noun*)	a building on Earth that sends and receives data from space
launched (*verb*)	sent into the sky with force
logging (*noun*)	the cutting down of trees for wood
meteorologists (*noun*)	scientists who study the air around Earth to predict the weather
navigate (*verb*)	to plan a path from one place to another
orbits (*verb*)	circles around a planet or star
radio signals (*noun*)	signals sent using a kind of wave that can be picked up by radios

sensors (*noun*) devices that measure where and how much of something there is, such as heat

space engineers (*noun*) people who design and build spacecraft, including satellites

USSR (*noun*) Union of Soviet Socialist Republics: a former country that was made up of Russia and several smaller countries

Index